ABDENOUR LAZEB

Optimization in Cloud Computing

ABDENOUR LAZEB

Optimization in Cloud Computing

ScienciaScripts

Imprint

Cover image: www.ingimage.com

This book is a translation from the original published under ISBN 978-620-6-71634-1.

Publisher:
Sciencia Scripts
is a trademark of
Dodo Books Indian Ocean Ltd. and OmniScriptum S.R.L publishing group

120 High Road, East Finchley, London, N2 9ED, United Kingdom
Str. Armeneasca 28/1, office 1, Chisinau MD-2012, Republic of Moldova, Europe
Printed at: see last page
ISBN: 978-620-7-91015-1

TABLE OF CONTENTS

1.INTRODUCTION

The introduction highlights the growing importance of optimisation in the context of Cloud Computing. It highlights the positive impact of optimisation on the performance, cost and security of cloud-based IT systems. In addition, it highlights the rapid evolution and challenges of the field of optimisation in the context of Cloud Computing, which amply justifies the need to develop effective strategies for optimising Cloud resources.

In an increasingly interconnected world, where organisations are increasingly reliant on cloud-based services and infrastructures, optimisation is becoming a key priority. This is because it enables the resources and capabilities offered by the cloud to be exploited to the full, while minimising costs and guaranteeing optimum performance.

Optimisation plays a key role in improving the performance of cloud-based IT systems. By identifying and resolving potential bottlenecks, it helps to optimise the use of resources and ensure a rapid response to user requests. It also helps to reduce waiting times and improve quality of service.

On the financial side, optimisation also offers significant benefits. By minimising the costs associated with cloud infrastructures, it enables businesses to make substantial savings. What's more, by optimising the use of resources, it avoids waste and encourages more efficient use of existing resources.

Security is also an essential aspect of optimisation in the context of Cloud Computing. By optimising configuration and security parameters, it is possible to reduce the risks associated with protection of sensitive data. In addition, an appropriate optimisation strategy can guarantee the confidentiality, integrity and availability of information stored in the cloud. However, it should be noted that the field of optimisation in the context of Cloud Computing is constantly evolving. Rapid technological advances and the new challenges posed by the growing needs of users require the development of optimisation strategies that are ever more effective and adapted to these new challenges.In conclusion, optimisation is essential in the context of Cloud Computing. It offers multiple advantages in terms of performance, cost and security. It is vital to develop effective optimisation strategies to take full advantage of the resources offered by the cloud. Only those who know how to optimise their systems judiciously will be able to remain competitive in a world that is increasingly connected and dependent on cloud-based services.

With this in mind, it is crucial to note the importance of continuous innovation and adaptation to changes in Cloud Computing. Optimisation strategies need to be constantly evolving to meet changing user needs and emerging technological challenges. This means keeping abreast of the latest advances and best practices in Cloud Computing.In addition, it should be emphasised that collaboration and knowledge exchange between optimisation experts are essential to foster innovation and continuous improvement. By sharing best practices, lessons learned and innovative solutions, it is possible to accelerate the development of cutting-edge optimisation strategies and maximise the benefits offered by Cloud Computing.Finally, it is important to note that optimisation is not just limited to the technical aspects of Cloud Computing. It also requires an in-depth analysis of the specific needs of each organisation and a clear understanding of the objectives to be achieved. By tailoring optimisation strategies to the unique requirements of each user, it is possible to maximise benefits and achieve results In short, optimisation is a key element in improving performance, reducing costs and enhancing security in the context of Cloud Computing. Developing effective optimisation strategies, keeping abreast of the latest trends and promoting collaboration between experts are essential to maximising the benefits offered by Cloud Computing. Through judicious optimisation, organisations will be able to remain competitive in an environment that is increasingly connected and dependent on cloud-based services.

1.1. Context of optimisation in Cloud Computing

The context of optimisation in Cloud Computing is examined in detail, focusing on the specific requirements of Cloud environments in terms of optimisation. It underlines the crucial importance of optimisation in meeting the demands of massive data processing, dynamic resource flexibility and the absolute guarantee of information security. In addition, it provides an in-depth analysis of the multiple approaches, techniques and tools available for optimising cloud systems, highlighting the unique and complex challenges that cloud computing professionals constantly face and that require innovative and scalable solutions. Fundamentally, this study aims to encourage a deeper understanding of the fundamental mechanisms of optimisation in the context of Cloud Computing, while maintaining conceptual clarity and practical relevance for successful implementation. It is essential to note that Continuous improvement in the optimisation of Cloud systems is crucial to ensure optimum performance and increased efficiency. As such, further research and exploration of innovative optimisation techniques is required to meet the evolving needs of Cloud Computing. In addition, it should be emphasised that optimisation in Cloud Computing has many potential benefits, such as cost savings, better resource utilisation and reduced latency, making it an ever-expanding area of interest and research. In conclusion, understanding and applying optimisation in Cloud Computing is essential to maximise the performance, efficiency and benefits of this revolutionary technology.

2.FOUNDATIONS OF CLOUD COMPUTING

Cloud Computing is a revolutionary computing model that gives users easy access at any time to a shared pool of configurable computing resources. Thanks to this innovative technology, users can benefit from high-performance networks, powerful servers, secure storage, advanced applications and a wide range of services. The main outstanding advantage of Cloud Computing is its extraordinary ability to deliver resources quickly and efficiently, without the need for complex management or constant interaction with the service provider.

There are several remarkable deployment models for Cloud Computing, each firmly and specifically tailored to specific needs. The revolutionary public cloud provides access to shared resources managed by an external service provider, offering great flexibility and remarkably low costs. The private cloud, on the other hand, is an infrastructure reserved exclusively for a specific organisation, offering an unrivalled level of control and security. Finally, the extraordinarily innovative hybrid cloud combines the advantages of the two previous models by enabling a seamless integration of unparalleled excellence between a public cloud and a private cloud.

The undeniably essential key concepts of Cloud Computing include on-demand self-service, which allows users to configure and manage their own resources without any external intervention. Broad, wide-area network access guarantees optimum connectivity that is unquestionably reliable for users, regardless of their geographical location. Resource pooling enables IT resources to be shared efficiently and intelligently between multiple users, resulting in unrivalled utilisation optimisation and significantly reduced costs. Incredibly fast elasticity allows users to dynamically adjust their resources to meet fluctuating needs, ensuring exceptionally optimal performance at all times. Finally, the precise measurement of services means that resource usage can be monitored and invoiced with impeccable accuracy, offering absolute transparency and financial management of unrivalled excellence.

In conclusion, Cloud Computing is an innovative technology that is revolutionising the world of computing by providing convenient, flexible and cost-effective access to IT resources in an undeniably remarkable way. With its extraordinarily high quality deployment models and key concepts of paramount importance, Cloud Computing offers many incredible and undeniable advantages to organisations and users, both technically and financially. By adopting this extraordinary technology, businesses can substantially improve their agility, competitiveness and capacity for innovation, while dramatically reducing costs and incredibly simplifying their IT management.

2.1. Definitions, concepts and detailed explanations

In the context of Cloud Computing, a number of specific terms need to be defined for better understanding. For example, Software as a Service (SaaS) offers software applications hosted in the cloud and accessible via a web browser to meet the growing needs of users. Thanks to this technology, businesses can now benefit from increased flexibility and reduced costs. They no longer need to manage a physical infrastructure and can simply concentrate on using the software without worrying about updating it, maintaining it or to its security. On the other hand, Platform as a Service (PaaS) provides a platform that enables customers to develop, run and manage applications without the complexity of building and maintaining the associated infrastructure. This enables businesses to concentrate on developing their applications and benefit from a rapid, secure development and deployment environment. As a result, they can save valuable time and resources. Finally, Infrastructure as a Service (IaaS) offers virtualised IT resources on demand, such as virtual machines, storage and networks. This enables businesses to obtain the resources they need quickly and easily, without having to invest in expensive hardware. What's more, with IaaS, businesses can easily adapt their resources according to their fluctuating needs, contributing to greater operational efficiency and significant cost savings.In conclusion, Cloud Computing offers several options to meet the needs of today's businesses. Whether it's SaaS, PaaS or IaaS, businesses can benefit from increased flexibility, reduced costs and greater operational efficiency. It It is therefore essential to understand these terms and their advantages in order to make the most of this constantly evolving technology.

3.TYPES OF CLOUD SERVICES

Cloud services fall into three main categories: IaaS (Infrastructure as a Service), PaaS (Platform as a Service) and SaaS (Software as a Service). Each of these services offers a different level of abstraction, allowing users to choose the degree of control and responsibility they wish to delegate to the Cloud service provider. Understanding the differences between these models is crucial to choosing the most appropriate solution for your needs and optimising the use of Cloud resources.

3.1. IaaS, PaaS, SaaS

IaaS (Infrastructure as a Service) provides a virtual computing environment, storage and network services via the Internet. PaaS (Platform as a Service) offers a platform and tools for developing, testing and deploying applications, while SaaS (Software as a Service) provides access to software and applications hosted in the Cloud. Each service eliminates the need to manage the underlying infrastructure, but offers different levels of control and customisation for end users. The use of these Cloud-based services has increased dramatically in recent years due to their many benefits for businesses. By opting for these service models, businesses can make significant cost savings by avoiding the need to invest in expensive physical infrastructures. What's more, it allows companies to concentrate on their core business without having to worry about managing IT resources. Thanks to IaaS, businesses can easily increase or decrease their computing capacity according to their needs, enabling them to optimise the use of resources and reduce costs PaaS offers developers a complete environment for develop, test and deploy applications quickly and efficiently. This allows them to concentrate on developing the application itself rather than worrying about the underlying infrastructure. With PaaS, developers can benefit from advanced features such as automatic resource management, transparent scalability and enhanced security. This high level of abstraction saves developers time and effort when developing new applications. By simplifying the deployment and management process, PaaS also enables development teams to work more efficiently, resulting in shorter time-to-market.

SaaS allows users to access software and applications hosted in the Cloud without having to install them locally. This makes it easy to access applications from any Internet-connected device. Users can simply connect to their SaaS account and access the software they need, without worrying about updates, security patches or hardware compatibility. Businesses can also benefit from flexible pricing models with SaaS, paying only for the services and functionality they really need.

In conclusion, Cloud-based services such as IaaS, PaaS and SaaS offer greater flexibility, scalability and efficiency for businesses, making them attractive options for organisations of all sizes. Thanks to these services, businesses can improve their agility, reduce costs and focus on their core business. Whether launching new applications, extending their infrastructure or adopting software solutions, Cloud computing offers a robust and scalable infrastructure to meet the needs of every business. By harnessing the benefits of IaaS, PaaS and SaaS, businesses can leverage the power and flexibility of the Cloud to drive growth and success in today's competitive marketplace.

4.OPTIMISATION NEEDS IN THE CLOUD

The need for optimisation in the cloud is becoming ever more pressing, particularly when it comes to scalability and elasticity. With ever-increasing requirements for storage and computing power, businesses need cloud solutions that can adapt quickly to these changes. The ability to scale resources according to demand is a key factor in guaranteeing optimum performance while keeping costs under control.

4.1. Scalability and elasticity

Scalability and elasticity are two essential characteristics in the field of cloud optimisation. Scalability, which refers to the ability to add new resources in response to an increase in workload, can be of vital importance in ensuring continuous, optimal performance. It means that available resources can be adapted quickly and easily to meet requirements, guaranteeing smooth and efficient evolution from infrastructure cloud infrastructure. Elasticity also plays a crucial role in cloud optimisation. By allowing resources to be dynamically reduced or increased according to demand, elasticity ensures efficient use of resources and the associated costs. When the workload decreases, resources can be reduced to avoid unnecessary wastage, saving valuable funds. Similarly, when workloads increase, resources can be rapidly increased to cope with growing demand, ensuring stable performance and performance. In short, scalability and elasticity combined offer a powerful means of optimising efficiency and performance in the cloud. They enable organisations to adapt quickly to changes in workloads, optimise the use of resources, and to reduce costs. By taking full advantage of these key features, businesses can benefit from a robust and agile cloud infrastructure, capable of responding to the challenges and demands of an ever-changing market.

5.KEY TECHNOLOGIES

Virtualisation is a key technology for optimising Cloud Computing, enabling virtual machines to be created from a single piece of physical hardware. Using hypervisors, this technology isolates applications and operating systems, optimising the use of hardware resources. Virtualisation contributes to flexibility and cost reduction, as it enables better use of servers by consolidating several virtual machines on a single physical server.

5.1. Virtualisation

Virtualisation is at the heart of Cloud Computing optimisation, offering an innovative approach to creating, managing and running virtual environments. Using virtualisation, users can create virtual instances that run applications and workloads efficiently, enabling better use of hardware resources and greater flexibility.

The creation of virtual environments provides businesses with a means of optimising their operations by creating pools of shared resources. This enables companies to consolidate their physical servers, reducing the number of servers required, while guaranteeing optimum performance. Thanks to virtualisation, businesses can also reduce the costs associated with purchasing and maintaining IT hardware, as they can make the most of existing resources. Virtualisation also makes it easier to migrate workloads between servers.

Companies can easily move their applications and their data from one server to another, contributing to greater operational efficiency and better use of resources. It also enables capacity to be quickly adjusted to meet changing needs, providing essential flexibility to meet increasing or decreasing demands.

Virtualisation plays a crucial role in reducing costs for businesses using Cloud Computing. By consolidating physical resources, businesses can reduce expenditure on IT infrastructure, energy and cooling. What's more, by optimising the use of resources, businesses can save on management and maintenance costs. Virtualisation also optimises performance by distributing the workload evenly across the available servers, ensuring a smooth and responsive user experience.

In conclusion, virtualisation is an essential element of optimisation in Cloud Computing. It enables the creation of virtual environments that offer better use of hardware resources, greater flexibility and lower costs. Thanks to virtualisation, businesses can optimise their operations, easily migrate workloads and improve performance, all contributing to their success in the Cloud.

6.OPTIMISATION METHODS

The section on optimisation methods explores different approaches to improving the efficiency and performance of Cloud systems. This includes the use of heuristic algorithms, resource optimisation techniques, cost reduction and performance improvement. These methods are essential to maximise the benefits of Cloud Computing and meet the evolving needs of users and applications.

6.1. Heuristic algorithms

In the subsection on heuristic algorithms, we look in detail at the various heuristic-based optimisation techniques. These approaches use rules of thumb and exploration strategies to find solutions of acceptable quality in a time reasonable time. We first look at local search algorithms, which focus on continuously improving an initial solution by exploring its neighbourhood. This involves performing a series of iterations to find the best solution.

Next, we study taboo search methods, which implement a memory mechanism to avoid revisiting solutions that have already been explored. This avoids infinite loops and improves the efficiency of the algorithm. In addition, we are taking a closer look at genetic algorithms, which are inspired by the process of natural evolution and use genetic operators such as selection, crossover and mutation to optimise solutions. These algorithms are based on the notion of population, where an initial population of solutions is generated and evolves over generations to converge towards an optimal solution. Finally, we explore other heuristic techniques specifically adapted to optimisation problems in Cloud Computing, such as particle swarm algorithms and evolutionary algorithms. Particle swarm algorithms are inspired by the collective behaviour of birds or insects to find optimal solutions, while evolutionary algorithms simulate the process of biological evolution to progressively improve solutions.

We analyse how they work, their advantages and limitations, with a particular focus on their application in the Cloud. Computing. Through this in-depth exploration, we aim to provide a comprehensive understanding of the different heuristic optimisation techniques available to solve complex Cloud Computing problems. We hope to give readers a detailed overview of these techniques and help them choose the one that best suits their specific Cloud Computing needs.

7.OPTIMISING RESOURCES

Optimising resources in cloud computing is essential to maximising operational efficiency and reducing costs. This involves strategically managing the use of IT resources, such as storage, network and computing power, to meet the changing needs of workloads. Companies can use intelligent resource allocation tools and management policies to optimise resource use while maintaining high performance. In addition, resource optimisation helps to minimise waste and improve environmental sustainability through more efficient use of energy and infrastructure. This can result in significant long-term cost savings, as well as a reduction in environmental impact. For example, by optimising the use of resources, companies can reduce their carbon footprint by cutting energy consumption and avoiding unnecessary investment in additional infrastructure. In addition, by using intelligent management policies, businesses can prioritise certain critical workloads, guarantee high performance for end users and maintain high availability of services. All of this is essential to ensuring a positive user experience and customer loyalty. Ultimately, optimising resources in the cloud offers many important benefits that can help businesses stay competitive in today's fast-moving marketplace.

7.1. Virtual Machine Management

Virtual machine management is a key aspect of optimising resources in cloud computing. It involves the efficient deployment, monitoring, scaling and management of virtual machines to meet workload requirements in a cost-effective way. Virtual machine management strategies include scheduling virtual machines for efficient use of physical resources, consolidating virtual machines to reduce resource wastage and automating management tasks to optimise operations. By dynamically adjusting the resources allocated to virtual machines as needed, businesses can optimise performance, reduce costs and improve the flexibility of their cloud infrastructure. This enables them to cope with growing demand without compromising quality of service. In addition, efficient management of virtual machines helps to minimise downtime, improve availability and ensure optimum use of resources. Businesses can also benefit from optimising energy consumption by adapting the resources allocated to virtual machines in line with fluctuations in demand. By investing in advanced management tools, businesses can automate virtual machine management processes, reducing manual workload and increasing operational efficiency. In summary, VM management is a key area for ensuring maximum efficiency in cloud computing. Organisations that adopt advanced management strategies can benefit from significantly improved performance, reduced costs and greater flexibility.

8.COST OPTIMISATION

Optimising costs in Cloud Computing involves minimising expenditure on infrastructure and Cloud services. This can be achieved by using pricing models that are well adapted to the specific needs of the business, by closely monitoring the consumption of resources to avoid waste, automating cost management processes, putting in place mechanisms to limit expenditure, and adopting best practice to reduce costs. For example, the use of reserved computing services or pre-emptive instances can lead to significant savings. In addition, working with cloud service providers to negotiate favourable rates, volume discounts and flexible pricing options can also help to optimise costs in the long term. In addition, implementing rigorous governance of cloud resources, eliminating unused or obsolete resources, and optimising application performance can also help reduce costs. Finally, strategic cloud spend planning, ongoing cost monitoring and regular spend analysis are essential to making informed decisions and maintaining sustainable cost optimisation in cloud computing.

8.1. Pricing Models

Cloud Computing pricing models vary depending on the services used, such as Infrastructure as a Service (IaaS), Platform as a Service (PaaS) and Software as a Service (SaaS). The most common models include pay-as-you-go, instance reservation, subscription pricing and hybrid models. It is essential for businesses to carefully evaluate the pricing models offered by cloud service providers to choose the one that best suits their needs in terms of flexibility, cost predictability and scalability, in order to optimise their cloud spend effectively. Understanding the different pricing models enables businesses to make informed decisions and maximise the value of their cloud investment. With pay-as-you-go pricing, businesses only pay for the IT resources they actually use, enabling them to reduce costs if usage is variable. Reserving instances, o n the other hand, allows companies to companies to benefit from reduced rates by reserving computing capacity in advance. Subscription pricing offers cost predictability by charging a fixed amount each month, regardless of actual usage. Finally, hybrid models combine different pricing models to meet specific business needs. By carefully evaluating these pricing models and choosing the most appropriate one, businesses can optimise their cloud spend and make the most of the services offered by cloud service providers.

9.PERFORMANCE OPTIMISATION

Optimising performance in Cloud Computing is essential to guarantee an optimal user experience. This means optimising latency, throughput and data processing speed. Companies are looking to minimise latency, i.e. the communication time between servers and end users, to improve the responsiveness of applications. Similarly, optimising throughput aims to maximise the amount of data that can be processed in a given period of time. These aspects are crucial for real-time services, such as video streaming or online gaming applications.

9.1. Latency and Throughput

Latency and throughput are key performance indicators in Cloud Computing. Latency refers to the time taken for a packet of data to traverse the network, while throughput refers to the amount of data that can be transferred in a given time interval. To optimise latency, companies are relying on distributed data centres to bring data closer to end users. In addition, the use of high-speed networks and caching technologies helps to reduce latency. To improve throughput, companies are investing in horizontal and vertical scaling solutions, as well as data compression techniques to optimise transfer. Another approach to optimising latency is to implement intelligent routing algorithms that route data efficiently and minimise latency.

These algorithms analyse network characteristics such as bandwidth, congestion and response times to determine the best route for data. In addition, companies can use distributed cache servers to store the most frequently requested data close to end users, significantly reducing latency. In terms of throughput, businesses can take advantage of horizontal scaling, which involves adding more servers or resources to increase processing capacity and data transfer speed. In addition, vertical scaling can improve throughput by increasing the resources (such as memory or computing power) on a single server. This optimises data processing and transfer operations, resulting in increased throughput. To further optimise throughput, businesses can invest in data compression solutions. These techniques reduce the size of the data before it is transferred, which increases throughput by reducing the time needed to transfer the data. In addition, the use of effective compression techniques ensures efficient use of bandwidth, contributing to an overall improvement in throughput. In conclusion, to optimise performance in Cloud Computing, it is essential to consider both latency and throughput. By investing in solutions such as distributed data centres, high-speed networks, caching techniques, intelligent routing algorithms, horizontal and vertical scaling, and data compression techniques, businesses can significantly improve the performance of their Cloud computing services. These measures guarantee faster response times, faster data transmission and an improved user experience.

10. SECURITY AND CONFIDENTIALITY

Data security and confidentiality are major concerns in cloud computing. Organisations face risks such as data theft, denial of service attacks and leaks of sensitive information. Robust security measures such as data encryption, identity and access management, and continuous monitoring of suspicious activity are essential to protect user and corporate data.

10.1. Challenges and Solutions

The challenges of security and confidentiality in cloud computing are particularly crucial in the digital age. With the rapid evolution of technology and the explosion of cyber threats, it is imperative to comply with regulations such as the General Data Protection Regulation (GDPR). In addition, it is essential to take proactive measures to protect data from internal and external threats. To meet these challenges, there are several effective solutions. The first is the use of advanced firewalls, which can detect and block unauthorised access. Secondly, rigorous data governance policies should be implemented to control access to and use of sensitive information. It is also advisable to adopt state-of-the-art intrusion detection technologies, which can identify suspicious activity and take appropriate measures to neutralise it. Finally, it is vital to train staff in good IT security practices. Increased awareness of these issues will enable employees to recognise and report security incidents, thereby minimising potential risks. In addition, it is essential to promote a culture of security within the organisation, encouraging responsible behaviour and providing regular educational resources. In conclusion, the challenges of security and confidentiality in Cloud Computing are real, but they can be overcome with the right tools. the right measures and strategies in place. By taking measures such as using advanced firewalls, implementing data governance policies, adopting intrusion detection technologies and training staff, businesses can strengthen their security posture and effectively protect their data in the ever-changing digital world.

11. CASE STUDIES

In this section, we will look at several case studies of companies that have successfully optimised their operations using Cloud Computing. These case studies will highlight the challenges they faced, the solutions they implemented and the results they achieved. Each case study will serve as a practical example to illustrate the different optimisation strategies discussed in this work.

11.1. Real Applications

The real-world applications of optimisation in Cloud Computing are extremely varied and versatile, covering a wide range of application areas. In addition to workload optimisation, which can significantly improve business efficiency and productivity, there are many other applications that take advantage of the benefits of optimisation in Cloud Computing. For example, many businesses have used specific optimisation strategies to reduce costs and improve the profitability of their operations. By optimising processes and resources, these companies have been able to reduce unnecessary expenditure and increase their profit margins. What's more, optimisation in Cloud Computing is also helping to improve the performance and quality of the services offered by these companies. More concretely, some companies have used optimisation in Cloud Computing to optimise their supply chain and transport operations. By optimising routes and itineraries, these companies have been able to reduce delivery times, minimise transport costs and improve customer satisfaction. Other companies have used Cloud-based optimisation to optimise their production processes and manufacturing operations. By using advanced optimisation models, these companies have been able to maximise resource utilisation, minimise downtime and improve the overall efficiency of their production activities. Cloud-based optimisation for e-commerce companies is also worth mentioning. By optimising prices, promotions and product recommendations, these businesses can attract and retain customers, increase sales and maximise revenues. Finally, optimisation in Cloud Computing also offers optimisation opportunities for service and consultancy companies. By using advanced optimisation techniques, these businesses can improve resource planning, optimise schedules and maximise the efficiency of their day-to-day operations. In conclusion, the real-world applications of optimisation in Cloud Computing are vast and diverse. They touch on many areas of business and offer enormous potential for improving performance, reducing costs and maximising profitability. The examples presented here illustrate the positive impact of Cloud Computing optimisation on business operations and highlight the tangible benefits it can bring.

12. CONCLUSION AND OUTLOOK

In conclusion, optimisation in Cloud Computing is a constantly evolving field, offering solutions to improve the performance, security and efficiency of Cloud services. The challenges of managing resources, reducing costs and guaranteeing performance will continue to drive research and development in this area. Future prospects include advances in virtualisation technologies, more flexible pricing models and more sophisticated optimisation strategies. Staying on top of emerging trends is essential to meet the growing needs of businesses and end-users. Against this backdrop of Optimisation in Cloud Computing is constantly evolving, and plays a crucial role for companies seeking to remain competitive in the marketplace. Indeed, the advantages offered by Cloud Computing in terms of flexibility, availability and cost reduction are increasingly sought after by organisations of all sizes. However, to make the most of these advantages, it is essential to optimise Cloud resources and performance. One of the major challenges in optimising Cloud computing lies in the efficient management of resources. Indeed, optimal resource allocation is crucial to ensuring efficient and cost-effective use of the Cloud. This means actively monitoring the use of resources, identifying potential bottlenecks and taking the necessary steps to resolve them. Effective resource management not only improves performance, but also reduces costs by avoiding waste. Another major challenge is guaranteeing performance. Cloud users expect high availability and response times, regardless of the number of simultaneous requests. To ensure an optimal user experience, it is therefore essential to implement sophisticated optimisation strategies. These strategies can include the implementation of caches, caching mechanisms to reduce response times, or load balancing to distribute the load evenly over several servers. In addition to these challenges, optimisation in Cloud Computing also faces security issues. The data stored in the Cloud can be sensitive and needs to be adequately protected against cyber-attacks. To guarantee data security, it is essential to adopt security measures such as data encryption, real-time monitoring of suspicious activity and the implementation of backup mechanisms to reduce the risk of loss. Despite these challenges, the future prospects for optimisation in cloud computing are promising. Technological advances in virtualisation will enable more efficient use of resources, while more flexible pricing models will offer businesses options tailored to their needs. to their specific needs. What's more, more sophisticated optimisation strategies will further improve the performance and efficiency of Cloud services. So it's crucial to keep abreast of emerging trends in Cloud Computing optimisation. Businesses and end-users need to be aware of the new technologies and optimisation strategies that are emerging. By staying at the forefront of these advances, it is possible to remain competitive in the market by meeting the growing needs of end-users and taking advantage of all the benefits offered by Cloud Computing.

12.1. Future Trends in Cloud Optimisation

Future trends in Cloud optimisation focus on the integration of artificial intelligence and machine learning to optimise resource management, workload planning and decision-making processes. The automation of optimisation tasks, the implementation of more advanced data management solutions and the use of emerging technologies such as serverless computing and containerisation are also key trends. At the same time, the focus on environmental sustainability and carbon footprint reduction in Cloud Computing will increase, with initiatives to integrate eco-responsible practices into optimisation strategies. When it comes to the integration of artificial intelligence and machine learning, businesses are looking for ways to leverage these technologies to improve decision-making and optimise Cloud performance. This will take the form of advanced machine learning models that can analyse large amounts of data to identify trends and patterns, enabling businesses to optimise their resource management and workload planning processes. Automation of optimisation tasks is another key area to watch. Companies are looking for ways to reduce the complexity and manual workload of cloud optimisation by Automating certain tasks. This can include the use of intelligent software robots that can perform repetitive and tedious tasks faster and more efficiently than humans. Implementing more advanced data management solutions is also an important aspect of cloud optimisation. Businesses are looking to improve the collection, storage and analysis of data to extract valuable insights and make more informed decisions. This can involve using technologies such as Big Data and the Internet of Things to collect data in real time and analyse it using advanced analytics tools. In addition to these key trends, there are also emerging technologies that are likely to play an important role in cloud optimisation in the future.

Serverless computing is one such technology, allowing businesses to develop and deploy applications without having to manage the underlying infrastructure. This can reduce costs and improve efficiency by allowing businesses to focus on application development rather than infrastructure maintenance. Containerisation is another emerging technology that enables businesses to build and deploy applications in a consistent and portable way, making it easier to migrate applications to the Cloud. Finally, the focus on environmental sustainability and reducing the carbon footprint in Cloud Computing will increase. Companies are looking for ways to reduce their environmental impact by adopting eco-responsible practices. This can take the form of using renewable energy sources to power data centres, optimising the energy efficiency of IT infrastructures and reducing electronic waste. Initiatives to integrate these eco-responsible practices into Cloud optimisation strategies will become increasingly important as awareness of the climate emergency continues to grow. In summary, future trends in Cloud optimisation will focus on the integration of artificial intelligence and machine learning, the automation o f optimisation tasks, the implementation of "cloud computing", and the development of "cloud computing". implementation of more advanced data management solutions and the use of emerging technologies such as serverless computing and containerisation. At the same time, the focus on environmental sustainability and reducing the carbon footprint of cloud computing will increase. These trends are essential for businesses looking to make the most of their cloud resources and remain competitive in the marketplace.

13.AUTOMATION AND ORCHESTRATION IN THE CLOUD

13.1. Importance of Automation

Automation in Cloud Computing can reduce operational complexity, reduce human error and improve efficiency. By using scripts and automation tools, businesses can manage resources more efficiently and reduce operational costs.

13.2. Automation tools

Popular automation tools include Ansible, Puppet and Chef, which enable configurations to be managed and deployed consistently. These tools also facilitate Infrastructure as Code (IaC) management, enabling repeatable and audited deployments.

13.3. Cloud Services Orchestration

Orchestration goes beyond automation by coordinating multiple automated tasks. Platforms like Kubernetes enable large-scale orchestration of containers, ensuring high availability and automatic scaling of applications.

14.HYBRID CLOUD AND MULTI-CLOUD

14.1. Advantages of the Hybrid Cloud

The hybrid cloud combines the advantages of the public and private clouds, offering flexibility, security and cost savings. Businesses can keep sensitive data private while using public resources for ephemeral needs or fluctuating workloads.

14.2. Multi-Cloud strategies

Adopting a multi-cloud strategy means avoiding dependency on a single supplier, improving resilience and optimising costs by choosing the best service for each specific need. Businesses need to develop the skills to manage multiple cloud environments effectively.

14.3. Multi-Cloud challenges

Identity management, security and data migration are major challenges in a multi-cloud environment. Companies need to invest in multi-cloud management solutions to monitor and secure applications across different providers.

15.ADVANCED CLOUD SECURITY

15.1. Zero Trust Security

The Zero Trust security model is based on the principle of never trusting an entity by default, but always verifying. This includes multi-factor authentication, network segmentation and strict access control policies.

15.2. Data Encryption

Encrypting data in transit and at rest is essential to protect sensitive information. Companies must use robust encryption keys and key management policies to ensure data confidentiality.

15.3. Continuous Monitoring and Incident Response

Continuous monitoring of cloud environments enables security incidents to be detected and responded to quickly. Companies need to set up security operations centres (SOCs) and use behavioural analysis tools to identify potential threats.

16.ENVIRONMENTAL IMPACT OF CLOUD COMPUTING

16.1. Energy consumption

Data centres consume large amounts of energy. Companies need to invest in energy-efficient technologies and sustainable management practices to reduce their carbon footprint.

16.2. Green Initiatives

Initiatives such as the use of renewable energy sources and optimising the energy efficiency of infrastructures can help to minimise the environmental impact of cloud computing.

16.3. Certifications and Compliance

Obtain certifications such as LEED (Leadership in Energy and Environmental Design) and follow ISO 14001 standards for management can demonstrate a company's commitment to sustainability.

17.THE FUTURE OF CLOUD COMPUTING

17.1. Artificial Intelligence and Machine Learning The integration of artificial intelligence and machine learning into cloud computing will continue to grow, offering advanced analysis, automated decision-making and resource optimisation capabilities.

17.2. Quantum Computing

Although still in its infancy, quantum computing promises to revolutionise cloud computing by solving complex problems much faster than traditional computers.

17.3. Regulatory developments

Data protection and privacy regulations will continue to evolve, affecting the way cloud services are deployed and managed. Businesses need to keep up to date with these developments to ensure compliance.

18.ADDITIONAL CASE STUDIES

18.1. Optimisation in the Healthcare Sector

A hospital has used cloud solutions to optimise the management of medical records, improve communication between departments and reduce waiting times for patients.

18.2. Financial Industry

A bank has adopted hybrid cloud strategies to guarantee the security of financial transactions while using public cloud services to analyse large quantities of transactional data in real time.

18.3. Education

A university has migrated to the cloud to offer e-learning services, enabling students to access educational resources from anywhere, while optimising costs and improving availability.

19.CLOUD MIGRATION STRATEGIES

19.1. Preparation assessment

Before migrating to the cloud, businesses need to assess their readiness in terms of skills, infrastructure and processes. This assessment helps to identify gaps and develop a detailed migration plan.

19.2. Migration Models

There are several models for migrating to the cloud, such as lift-and-shift migration, re-architecture and replacement. Each model has its advantages and disadvantages, and the choice depends on the company's objectives and constraints.

19.3. Risk Management

Risk management is crucial when migrating to the cloud. Businesses need to identify potential risks, draw up plans and ensure business continuity during the migration.

20.GOVERNANCE AND COMPLIANCE IN THE CLOUD

20.1. Governance Policies

Implementing robust governance policies helps to ensure that cloud resources are used responsibly and aligned with business objectives. This includes access management, regulatory compliance and ongoing monitoring.

20.2. Regulatory Compliance

Businesses need to comply with various regulations, such as GDPR, HIPAA and SOX, when using cloud services. This means putting in place appropriate controls to protect sensitive data and guarantee the confidentiality and integrity of information.

20.3. Audits and Reports

Performing regular audits and generating compliance reports are essential to maintaining transparency and accountability in the use of cloud services. Companies should use automated audit tools to simplify this process.

21.ECONOMIC IMPACT OF CLOUD COMPUTING

21.1. Cost reduction

Cloud computing enables businesses to cut costs by eliminating the need to invest in expensive infrastructure and offering flexible pricing models.

21.2. Economies of scale

Cloud service providers benefit from economies of scale, enabling them to offer services at competitive rates. Companies can take advantage of these savings to optimise their IT budgets.

21.3. Return on Investment (ROI)

Calculating ROI is essential for assessing the economic benefits of moving to the cloud. Companies need to analyse initial costs, savings and productivity gains to determine ROI.

22.ARTIFICIAL INTELLIGENCE APPLICATIONS IN THE CLOUD

22.1. Data Analysis

Cloud computing provides an ideal platform for analysing large amounts of data using artificial intelligence algorithms. Businesses can use these capabilities to gain valuable insights and make data-driven decisions.

22.2. Intelligent Automation

Artificial intelligence can be used to automate complex tasks such as customer service management, predictive maintenance and supply chain optimisation. The cloud provides the infrastructure needed to deploy these solutions on a large scale.

22.3. Machine learning

Cloud platforms offer tools and services for developing, training and deploying machine learning models. This enables businesses to create intelligent applications that can learn and adapt based on data.

23.OPTIMISING CLOUD APPLICATION PERFORMANCE

23.1. Caching

The use of local and distributed caches can reduce latency and improve the performance of cloud applications. Companies can implement caching solutions to temporarily store frequently accessed data.

23.2. Load sharing

Load balancing helps to distribute traffic between several servers, ensuring optimum use of resources and high availability of applications. Companies need to use load balancers to manage traffic efficiently.

23.3. Database optimisation

Database optimisation is essential for improving the performance of cloud applications. This includes appropriate indexing, query caching and the use of NoSQL databases for specific workloads.

24.CASE STUDIES AND CASE STUDIES

24.1. Enterprise Application Migration

A concrete example of the migration of an enterprise application to the cloud, including the challenges encountered, the solutions implemented and the results obtained.

24.2. Optimising a Data Management System

A case study on the optimisation of a data management system in the cloud, detailing the techniques used to improve performance and reduce costs.

24.3. Implementing AI in e-commerce

A case study in the use of artificial intelligence in e-commerce to personalise user experiences and increase sales.

APPENDICES

Appendix A: Glossary of Terms

A detailed glossary of key terms and concepts used in cloud computing, to help readers better understand the content of the document.

Appendix B: Schematics and diagrams

Schematics and diagrams illustrating the cloud architectures, workflows and optimisation processes described in the document.

Appendix C: Examples of Automation Scripts

Examples of automation scripts for managing cloud resources, showing how to use tools such as Ansible and Terraform.

Appendix D: Resources and further reading

A list of resources and further reading for those who want to learn more about optimisation in cloud computing.

25.DEPLOYMENT OF CLOUD SERVICES

25.1. Deployment Strategies

Deployment strategies include continuous deployment (CI/CD), blue-green deployment and canary deployment. Each strategy has its advantages in terms of minimising risk and ensuring uninterrupted commissioning.

25.2. Deployment Tools

Tools such as Jenkins, GitLab CI and Spinnaker are essential for automating deployment pipelines. These tools enable continuous integration and delivery, ensuring that new versions of the software are deployed quickly and securely.

25.3. Version Management

Version management is crucial to ensuring the compatibility and stability of deployed applications. Companies need to adopt semantic versioning practices and use version management tools to track and manage the different versions of applications.

26.PERFORMANCE MONITORING AND MANAGEMENT

26.1. Monitoring tools

Monitoring tools such as Prometheus, Grafana and Datadog enable you to track the performance of cloud services in real time. They offer detailed visualisations and alerts to identify and resolve problems quickly.

26.2. Log analysis

Log analysis is essential for diagnosing problems and optimising performance. Tools such as ELK Stack (Elasticsearch, Logstash, Kibana) help to collect, analyse and visualise application logs.

26.3. Incident Management

Incident management involves putting in place procedures to detect, diagnose and resolve problems quickly. Companies need to adopt incident management practices and use tools such as PagerDuty to coordinate incident response.

27.ADVANCED SECURITY AND CONFIDENTIALITY

27.1. Authentication and authorisation

Authentication and authorisation are the cornerstones of security in the cloud. Implementing Single Sign-On (SSO), OAuth, and Attribute Based Roles (ABAC) enables secure access to cloud resources.

27.2. API security

API security is crucial to protecting interactions between services. Companies must use API gateways, apply strict security policies, and monitor API calls to detect anomalies.

27.3. Identity and Access Management

Identity and Access Management (IAM) allows you to control who can access which resources. Companies need to adopt robust IAM policies and use tools such as AWS IAM, Azure AD and Google Cloud IAM to manage access.

28.COST OPTIMISATION

28.1. Cost Management Tools

Tools such as AWS Cost Explorer, Azure Cost Management and Google Cloud Billing enable costs to be monitored and optimised. They offer detailed insights into resource consumption and help to identify opportunities for cost reduction.

28.2. Licence optimisation

Optimising software licences is crucial to reducing expenditure. Companies need to monitor licence usage, eliminate unused licences and negotiate flexible licence agreements with suppliers.

28.3. Cost Reduction Strategies

Cost reduction strategies include using reserved instances, optimising workloads to use spot/preemptible instances, and automating the shutdown of unused resources.

29.ADDITIONAL CASE STUDIES

29.1. Health sector

A major hospital has migrated its medical records management systems to the cloud, improving access to data for doctors while reducing maintenance costs for local servers.

29.2. Banking sector

An international bank has adopted a multi-cloud strategy to ensure the resilience and availability of its financial services, while optimising costs by using different suppliers for different needs.

29.3. Education sector

A university has used cloud solutions to create a robust e-learning environment, enabling students to access courses and educational resources flexibly and efficiently.

30.DETAILED APPENDICES

Appendix E: Full Case Studies

Comprehensive case studies, detailing the migration process, the challenges faced, the solutions implemented and the results achieved, for various sectors such as health, finance, education and more.

Appendix F: Explanatory diagrams and schematics

Detailed diagrams illustrating cloud architectures, automation workflows, deployment strategies and security configurations.

Appendix G: Examples of scripts

Sample scripts for deployment automation, resource management and performance monitoring, using tools such as Terraform, Ansible and Jenkins.

Appendix H: Comparison tables

Comparative tables of different cloud service providers, cost management tools and security solutions, to help businesses choose the best options for their needs.

Appendix I: Glossary of Advanced Terms

A glossary of advanced cloud computing terms and concepts, to help readers better understand the technical aspects covered in the document.

Appendix J: Educational resources

A list of online courses, books and research articles for those who want to learn more about cloud computing and its optimisation

REFERENCE

1. Ullah, F., et al (2023). *Cloud Computing Optimization: Current Trends and Future Directions*. IEEE Access, 11, 14725-14740. [doi:10.1109/ACCESS.2023.3248765](https://doi.org/10.1109/ACCESS.2023.3248765)

2. Ali, O., Shrestha, A., & Ijaz, A. (2022). *Comprehensive Review of Cloud Computing Foundations: Architecture, Applications, and Trends*. Future Internet, 14(5), 139. [doi:10.3390/fi14050139](https://doi.org/10.3390/fi14050139)

3. Kumar, S., et al (2023). *Service Models and Deployment Strategies in Cloud Computing: A Survey*. Journal of Cloud Computing, 12(1), 25 [doi:10.1186/s13677-023-00294-6](https://doi.org/10.1186/s13677- 023-00294-6)

4. Zhang, Y., Chen, L., & Zhang, S. (2023). *Scalability and Elasticity in Cloud Computing: A Survey of Algorithms and Metrics. ACM Computing Surveys, 56(2), 34.[doi:10.1145/3582391](https://doi.org/10.1145/3582391)

5. Mishra, S., et al (2023). *Advances in Virtualization for Cloud Computing: Trends and Challenges. Journal of Cloud Computing: Advances, Systems and Applications, 12(1), 47 [doi:10.1186/s13677- 023-00299-1](https://doi.org/10.1186/s13677-023-00299-1)

6.Li, J., et al (2023). *Heuristic and Metaheuristic Optimization Techniques in Cloud Computing: A Review*. Journal of Parallel and Distributed Computing, 175, 128-146. [doi:10.1016/j.jpdc.2023.01.013](https://doi.org/10.1016/j.jpdc.2023. 01.013)

7. Zhao, Y., & Tang, M. (2023). *Resource Allocation and Management in Cloud Computing: Techniques and Tools*. IEEE Transactions on Cloud Computing, 11(3), 789-801.

[doi:10.1109/TCC.2023.3250407](https://doi.org/10.1109/TCC.2023.3 250407)
8. Srirama, S. N., & Buyya, R. (2022). *Cost Optimization Strategies for Cloud Computing: A Comprehensive Review. Journal of Cloud Computing: Advances, Systems and Applications, 11(1), 54.[doi:10.1186/s13677-022-00289-5](https://doi.org/10.1186/s13677- 022-00289-5)

9. Rana, O., et al (2023). *Performance Optimization in Cloud Computing: Latency, Throughput, and Response Time. Future Generation Computer Systems, 141, 1-15.
[doi:10.1016/j.future.2023.01.009](https://doi.org/10.1016/j.future.20 23.01.009)

10. Sun, Y., et al (2023). *Security and Privacy in Cloud Computing: Challenges and Solutions. IEEE Transactions on Dependable and Secure Computing, 20(2), 645-658.
[doi:10.1109/TDSC.2022.3150271](https://doi.org/10.1109/TDSC.202 .3150271)

11.Ghaffari, A., et al (2023). *Case Studies in Cloud Computing Optimization: Practical Applications and Lessons Learned. Journal of Cloud Computing, 12(1), 69. [doi:10.1186/s13677-023-00301- y](https://doi.org/10.1186/s13677-023-00301-y)
12. Singh, S., & Choudhary, G. (2023). *Future Trends in Cloud Computing Optimization: AI and Machine Learning Approaches. ACM Computing Surveys, 56(3), 45.[doi:10.1145/3582484](https://doi.org/10.1145/3582484)

13. Kaur, H., et al (2024). *Emerging Trends in Cloud Computing Optimization: AI, Serverless, and Sustainability*. Journal of Cloud Computing,13(1),12.[doi:10.1186/s13677-024-00305-5](https://doi.org/10.1186/s13677-024-00305-5)

Printed by Books on Demand GmbH, Norderstedt / Germany